Energy Sector Standard of the People's Republic of China

NB/T 34064-2018

Specification for operation management of densified biofuel projects for heating boilers

生物质锅炉供热成型燃料工程运行管理规范

(English Translation)

China Water & Power Press
中国水利水电出版社
Beijing 2024

图书在版编目（CIP）数据

生物质锅炉供热成型燃料工程运行管理规范 : NB/T 34064-2018 = Specification for operation management of densified biofuel projects for heating boilers(NB/T 34064-2018) : 英文 / 国家能源局发布. -- 北京 : 中国水利水电出版社, 2024. 10.

ISBN 978-7-5226-2774-8

Ⅰ. TK227.1-65

中国国家版本馆CIP数据核字第202481HX06号

Energy Sector Standard of the People's Republic of China

中华人民共和国能源行业标准

Specification for operation management

of densified biofuel projects for heating boilers

生物质锅炉供热成型燃料工程运行管理规范

NB/T 34064-2018

(English Translation)

Issued by National Energy Administration of the People's Republic of China

国家能源局　发布

Translation organized by China Renewable Energy Engineering Institute

水电水利规划设计总院　组织翻译

Published by China Water & Power Press

中国水利水电出版社　出版发行

Tel: (+ 86 10) 68545888　68545874

sales@mwr.gov.cn

Account name: China Water & Power Press

Address: No.1, Yuyuantan Nanlu, Haidian District, Beijing 100038, China

http: //www.waterpub.com.cn

中国水利水电出版社微机排版中心　排版

北京中献拓方科技发展有限公司　印刷

210mm×297mm　16 开本　1.25 印张　50 千字

2024 年 10 月第 1 版　2024 年 10 月第 1 次印刷

Price（定价）: ¥ 200.00

About English Translation

This English version is one of China's energy sector standard series in English. Its translation was organized by China Renewable Energy Engineering Institute authorized by National Energy Administration of the People's Republic of China in compliance with relevant procedures and stipulations. This English version was issued by National Energy Administration of the People's Republic of China in Announcement [2023] No. 8 dated December 28, 2023.

This version was translated from the Chinese Standard NB/T 34064-2018, *Specification for operation management of densified biofuel projects for heating boilers*, published by China Water & Power Press. The copyright is reserved by National Energy Administration of the People's Republic of China. In the event of any discrepancy in the implementation, the Chinese version shall prevail.

Many thanks go to the staff from the relevant standard development organizations and those who have provided generous assistance in the translation and review process.

For further improvement of the English version, any comments and suggestions are welcome and should be addressed to:

China Renewable Energy Engineering Institute
No. 2 Beixiaojie, Liupukang, Xicheng District, Beijing 100120, China
Website: www.creei.cn

Translating organizations:

Academy of Agricultural Planning and Engineering, MARA

China Renewable Energy Engineering Institute

Translating staff:

MENG Haibo	FENG Jing	SHEN Xiuli	YE Bingnan
LI Lijie	XING Haohan	LIU Huan	CHEN Mingsong
LIU Xiangyang			

Review panel members:

QIAO Peng	POWERCHINA Northwest Engineering Corporation Limited
LI Zhongjie	POWERCHINA Northwest Engineering Corporation Limited
YAN Wenjun	Army Academy of Armored Forces, PLA
QIE Chunsheng	Senior English Translator
GUO Jie	POWERCHINA Beijing Engineering Corporation Limited
CHE Zhenying	IBF Technologies Co., Ltd.
CONG Hongbin	Academy of Agricultural Planning and Engineering, MARA

National Energy Administration of the People's Republic of China

翻译出版说明

本译本为国家能源局委托水电水利规划设计总院按照有关程序和规定，统一组织翻译的能源行业标准英文版系列译本之一。2023 年 12 月 28 日，国家能源局以 2023 年第 8 号公告予以公布。

本译本是根据中国水利水电出版社出版的《生物质锅炉供热成型燃料工程运行管理规范》NB/T 34064—2018 翻译的，著作权归国家能源局所有。在使用过程中，如出现异议，以中文版为准。

本译本在翻译和审核过程中，本标准编制单位及编制组有关成员给予了积极协助。

为不断提高本译本的质量，欢迎使用者提出意见和建议，并反馈给水电水利规划设计总院。

地址：北京市西城区六铺炕北小街 2 号
邮编：100120
网址：www.creei.cn

本译本翻译单位：农业农村部规划设计研究院
水电水利规划设计总院

本译本翻译人员：孟海波 冯 晶 沈秀丽 叶炳南
李丽洁 邢浩翰 刘 欢 陈明松
刘向阳

本译本审核人员：

乔 鹏 中国电建集团西北勘测设计研究院有限公司
李仲杰 中国电建集团西北勘测设计研究院有限公司
闫文军 中国人民解放军陆军装甲兵学院
郄春生 英语高级翻译
郭 洁 中国电建集团北京勘测设计研究院有限公司
车振英 一百分信息技术有限公司
丛宏斌 农业农村部规划设计研究院

国家能源局

Contents

Foreword

This standard is drafted in accordance with the rules given in the GB/T 1.1-2009 *Directives for Standardization—Part 1: Structure and drafting of standards*.

National Energy Administration of the People's Republic of China is in charge of the administration of this code. China Renewable Energy Engineering Institute has proposed this code, and is responsible for its routine management and the explanation of specific technical contents. Comments and suggestions in the implementation of this code should be addressed to:

China Renewable Energy Engineering Institute
No. 2 Beixiaojie, Liupukang, Xicheng District, Beijing 100120, China

Chief development organizations:

Academy of Agricultural Planning and Engineering, MARA

Beijing Yifang Sunshine Energy Technology Co., Ltd.

Chief drafting staff:

YAO Zonglu	ZHAO Lixin	MENG Haibo	HUO Lili
CONG Hongbin	YUAN Yanwen	ZHAO Kai	FENG Jing
LUO Juan	REN Yawei	LI Lijie	WANG Guan
DONG Yifang			

Specification for operation management of densified biofuel projects for heating boilers

1 Scope

This standard specifies the methods and requirements of the operation, maintenance, and safe operation of densified biofuel projects for heating boilers.

This standard is applicable to densified biofuel projects for heating boilers with an annual production capacity of 10000 t and above.

2 Normative references

The following documents are indispensable for the application of this document. For dated references, only the edition cited applies. For undated references, the latest version (including all amendments) applies.

GB 4387, *Safety regulation for railway and road transportation in plants of industrial enterprises*

GB/T 12801, *General principles for the requirements of safety and health in production process*

GB 15577, *Safety regulations for dust explosion prevention and protection*

GB/T 15605, *Guide for pressure venting of dust explosions*

GB 50601, *Code for construction and quality acceptance for lightning protection engineering of structures*

JB/T 10354, *Operating code for industrial boilers*

NY/T 1882, *Technical conditions for densified biofuel molding equipment*

NY/T 1883, *Testing method for densified biofuel molding equipment*

NY/T 1915, *Densified biofuel-Terminology and definitions*

3 General provisions

3.1 Operation management

3.1.1 The operating procedure shall be formulated to specify the requirements for raw materials and propose the main process technical indicators, and attached with the process flow and equipment list.

3.1.2 The operation and management manual shall be formulated to specify the duties of operators, post operating procedure, equipment operating procedure and maintenance procedure.

3.1.3 A work shift system shall be established to specify the checklist and responsibility boundary between work shifts.

3.1.4 The operation manager shall be familiar with the process and the operational requirements of equipment and facilities, operation techniques, performance indicators and emergency response measures.

3.1.5 The operators shall be familiar with the operational requirements, operation techniques and performance indicators of the facilities and equipment on his post, and be familiar with the process flow, and observe the post operating procedure.

3.1.6 The operation and maintenance personnel shall receive specialized technical training.

3.1.7 The equipment operators shall fill in the operation record in an accurate and timely manner. The operation manager shall inspect and verify the original records on a regular basis.

3.1.8 The operation manager and operators shall inspect the operation of equipment, electrical systems, and meters according to management and process requirements, and fill in the inspection records.

3.1.9 Necessary operation and safety signs shall be provided at the obvious positions for the control room and facilities and equipment.

3.1.10 Electrical operation and maintenance shall be in strict accordance with safe operation specifications. The operation and maintenance of electrical equipment shall be in strict accordance with equipment operating procedure. The electrical safety shall be checked on a regular basis, including the operation status of explosion-proof motors and switches.

3.1.11 Cleaning shall be conducted on a regular basis to keep the facilities and equipment clean and tidy.

3.1.12 Transportation management in the plant area shall comply with GB 4387.

3.2 Maintenance

3.2.1 A comprehensive equipment maintenance plan shall be formulated, which mainly includes equipment name, maintenance items, maintenance interval, maintenance staff, etc., and maintenance records shall be made.

3.2.2 A three-level maintenance system for equipment shall be established, including daily maintenance, level 1 maintenance, and level 2 maintenance.

3.2.3 Maintenance staff shall strictly implement the maintenance plan and inspection and acceptance systems. Regular maintenance shall be conducted to the facilities, equipment, meters, pipelines, supports and hangers, cover plates, platforms, etc. on the production line.

3.2.4 The maintenance equipment, spare parts and tools shall be provided and properly kept.

3.3 Safe operation

3.3.1 The safety protection measures, safety operating procedure and fire emergency plans shall be formulated as per GB/T 12801, taking into account the production characteristics. Life-saving facilities and supplies shall be provided.

3.3.2 Emergency preparedness plans shall be formulated for fire, flammables and explosives, and leakage of harmful gases, etc., and safety signs and safety colors shall be set up in the locations with safety hazards.

3.3.3 Emergency preparedness plans shall be formulated for natural disasters such as floods, hail, earthquakes.

3.3.4 The operation management of explosion-proof, explosion resistance and explosion venting equipment or facilities shall comply with GB 15577 and GB/T 15605 and other relevant regulations.

3.3.5 Lightning protection and earthing, explosion-proof testing and maintenance of buildings (structures) shall comply with GB 50601 and other relevant regulations.

3.3.6 A regular safety education system shall be established. Operators shall be provided with systematic safety education, and safety training on protection against fire, water, lightning, and electric shock. Regular fire drills shall be organized.

3.3.7 Workers specialized in electrical, welding, boiler, inspection and calibration, etc., shall receive professional skills training and safety technical training.

3.3.8 Operation manager, operators, and maintenance personnel shall understand the potential hazards and harmful factors associated with their posts, and master the corresponding countermeasures.

3.3.9 Operators of each post shall wear personal protective equipment (PPE) during work and follow safety procedures.

3.3.10 When maintaining mechanical equipment, temporary power lines shall not be connected without permission. During maintenance, the caution board indicating maintenance in progress shall be hung.

3.3.11 Welding, gas cutting or other open-flame operations are strictly prohibited in the raw material storage area and finished product warehouse.

4 Raw material measurement and storage

4.1 Operation management

4.1.1 Measurement record system shall be implemented on the raw materials delivered to the plant. The measurement record shall include the vehicle number, transporter, arrival time, leave time, and the type, source, quality, moisture content, etc. of raw materials.

4.1.2 The raw material measurement system shall be kept in good condition, and the devices in the system shall be in normal use.

4.1.3 The operator shall regularly inspect the measurement system, and keep the calibration record.

4.1.4 The operator shall make daily receipt and storage records of raw materials delivered to the plant, monthly statistical reports and data backup.

4.1.5 The operator shall make a record of his work on duty and shift process.

4.1.6 When the measurement system fails, a backup system or manual recording method shall be adopted to ensure that the measurement work is implemented normally and the records are kept complete and accurate.

4.1.7 The operator shall conduct a preliminary test on the moisture content and impurities of the raw materials delivered to the plant according to the production process requirements. Noncompliant raw materials shall be rejected.

4.1.8 The raw materials shall be stacked and partitioned according to categories, and fire lanes shall be set up. The stacks shall be protected from rain, snow, wind, and dust.

4.2 Maintenance

4.2.1 The operator shall regularly inspect and maintain the measurement system.

4.2.2 The measurement equipment shall be verified by the national metrology institute.

4.3 Safe operation

4.3.1 Indicating signs shall be set up in front of and behind the measurement equipment.

4.3.2 A speed bump shall be installed at the place 10 m approaching the measurement equipment.

4.3.3 The lightning protection and earthing facilities of the measurement equipment shall be kept in good conditions.

4.3.4 The raw material storage area shall be inspected regularly to keep the fire lanes clear and the firefighting facilities complete and effective.

5 Crushing

5.1 Operation management

5.1.1 The operator shall strictly follow the user's manual and operating procedure of the crusher.

5.1.2 Before crushing, the raw materials shall be inspected and cleared of metal, gravels, etc.

5.1.3 Before starting the crusher, the operator shall check that the seat lock bolts, the screen mesh, the protective cover and the operating door are securely locked.

5.1.4 After startup, the crusher shall operate idly for at least 1 min before feeding; the operator shall pay attention to the working current and sound of the equipment from time to time.

5.1.5 Before shutdown, the crusher shall keep running idly for at least 2 min after the front equipment in the process chain stops and the residual materials are crushed and drawn out.

5.1.6 No material shall be left in the machine after shutdown, and the site and equipment shall be cleaned.

5.2 Maintenance

5.2.1 The equipment shall be lubricated and maintained on a regular basis according to the user's manual and operating procedure.

5.2.2 The operation condition of the equipment shall be regularly inspected. If any hazards identified, the operation shall be suspended and the hazards be addressed as soon as possible.

5.2.3 Service and maintenance shall be conducted by professionals, and recorded in time; the main power supply shall be cut off and a board indicating maintenance in progress shall be hung at a conspicuous position before maintenance.

5.2.4 The operation managers shall supervise the maintenance work to prevent malpractice.

5.3 Safe operation

5.3.1 The operators not intended to the process flow are strictly prohibited from operating the related equipment.

5.3.2 The operator shall not be absent from work position once the machine starts, and shall stand by the side of the inlet when feeding, to avoid the occurrence of safety accidents caused by splashed materials or debris; it is strictly prohibited to use the hand or other hard objects to dispense materials at the inlet and outlet.

5.3.3 During the operation of the equipment or before its complete shutdown, it is strictly prohibited to open the operation door and to clean or repair the crusher.

5.3.4 If any anomaly occurs during operation, the crusher shall be stopped immediately, and troubleshot by the maintenance personnel, and shall not be put into operation until it resumes to normal operation; the operator shall not disassemble the equipment without permission.

6 Drying

6.1 Operation management

6.1.1 Before operation, the operator shall check the transmission systems, the tightening of components, the lubrication condition, the electrical meters, buttons, and indicators; and check the material blocking, and the blocking of guard mesh of the fan outlet. The operator shall make sure no unauthorized persons or obstacles around the device.

6.1.2 During operation, the operator shall check whether the surface temperature of the equipment and the temperature rise of the bearings are within the specified range, whether the air ducts are blocked and show signs of dust accumulation, whether the current and voltage readings are normal, whether the equipment is operating with abnormal sounds or odors, whether the temperature in the drying cylinder are normal, and whether the spark detector is operating normally.

6.1.3 The drying temperature shall be such that ensures that the moisture of the raw materials after drying meets the process requirements.

6.1.4 The machine shall not be shut down until all the materials are discharged.

6.1.5 After each shutdown of the drying machine, the induced draft fan shall not stop until the temperature drops to within the specified range.

6.1.6 The operation of the drying machine using the boiler shall comply with JB/T 10354.

6.2 Maintenance

6.2.1 The drying machine shall be maintained in accordance with 5.2.

6.2.2 The drying machine shall be cleared of impurities inside the drying machine in a timely manner.

6.2.3 The power supply shall be cut off during shutdown cleaning or maintenance, with a board indicating the cleaning or maintenance in progress. The temperature inside the drying machine shall drop to ambient temperature before cleaning or maintenance.

6.3 Safe operation

6.3.1 Fire accidents caused by electrical equipment, external open flames, etc. shall be strictly prevented during drying.

6.3.2 The drying temperature shall not exceed the upper limit of the working temperature of the equipment.

7 Compression and molding

7.1 Operation management

7.1.1 The operator shall strictly follow the user's manual and operating procedure of the molding machine.

7.1.2 Before startup, the operator shall check the fasteners on the feed port shield and the feeder.

7.1.3 During startup and operation, the operator shall pay close attention to the working current, and overload operation is strictly prohibited.

7.1.4 During operation, the inspection system shall be strictly implemented. The operator shall check the display of each instrument, and check for abnormal noise or vibration, and the feeding and discharging of materials.

7.1.5 The appearance quality of the finished products shall be checked from time to time, the moisture content and feed supply shall be adjusted in time, and the relevant equipment shall be adjusted if necessary.

7.1.6 The operator shall pay close attention to the operation of the equipment. If anomalies are identified, the equipment shall be shut down immediately, and restarted after troubleshooting.

7.1.7 The residual material in the molding machine shall be properly treated before shutdown.

7.1.8 After the operation is completed, the power supply shall be cut off, the site shall be cleared, and the equipment shall be checked for missing parts.

7.2 Maintenance

7.2.1 The key components shall be checked, adjusted and replaced according to the maintenance requirements of the molding equipment.

7.2.2 The transmission system, the lubrication system and the lubrication of parts shall be checked on a regular basis.

7.2.3 The operation manager shall supervise the equipment maintenance work to prevent malpractice.

7.2.4 The maintenance plan shall be proposed according to the wear of equipment, maintenance shall be implemented as schedule, and shall be recorded in time.

7.3 Safe operation

7.3.1 The operator shall wear tight-fitting overalls, fasten cuffs, and wear helmet and dust mask; the filter paper of the dust mask shall be replaced regularly.

7.3.2 During operation, it is strictly prohibited to open the guard shield, the handhole cover on the feeder, etc. or put hands or other objects into the pressing room, feeder or conditioner.

7.3.3 During shutdown check of the pressing room, the operator shall wear gloves to prevent burns.

7.3.4 During maintenance, the power supply must be cut off, and a board indicating maintenance in progress shall be hung.

8 Control system

8.1 Operation management

8.1.1 The operator shall inspect the electrical equipment and instruments regularly, and treat abnormalities in time.

8.1.2 The operator shall keep an eye on the control signals and maintain an operation log. When the fault occurs, the machine shall be shut down immediately and the maintenance personnel or operation manager shall be informed.

8.1.3 During the operation of the control system, when the circuit breaker trips off or blows out, the machine shall not be restarted until the cause is identified.

8.1.4 When a complete set of equipment or a machine in the production line fails, all equipment before the fault point shall be shut down, immediately inspected, and troubleshot.

8.2 Maintenance

8.2.1 Complete files shall be established for instruments and meters, and their components, sensors, transducers, indicators, etc. shall be inspected on a regular basis.

8.2.2 The components of the control equipment shall be kept clean and free of corrosion. Dial scales shall be legible, with intact nameplates, labels, and lead seals. The control room shall be kept clean and regularly inspected, and desiccants shall be replaced regularly.

8.2.3 It is strictly forbidden to use cleaning agents that might damage components.

8.2.4 Repair of instruments and meters shall be undertaken only by professionals. The maintenance of important instruments shall be performed by a specialized maintenance department or by the manufacturer, and the instruments shall not be disassembled without permission.

8.2.5 Instruments included in the national list of mandatory verification shall be regularly verified by relevant departments.

8.3 Safe operation

8.3.1 The equipment and instruments in the control room shall be operated by designated personnel, and unauthorized workers shall not be allowed to enter the control room.

8.3.2 The control instruments and meters shall be checked on a regular basis for measures against moisture, dust, lightning and static electricity.

8.3.3 Smooth and uninterrupted communication between the control room and each process shall be maintained.

9 Dedusting

9.1 Operation management

9.1.1 Before operation, the equipment shall be inspected in accordance with 7.1.1.

9.1.2 The dust conveying system shall be regularly inspected for leaks and wear. If any fault is identified, it should be troubleshot in time.

9.1.3 The dedusting status at the exhaust port of the dedusting system shall be regularly observed. If an anomaly is identified, it shall be troubleshot.

9.1.4 The operation of the dedusting equipment shall be stopped after the crushing and molding equipment is completely shut down.

9.1.5 After shutdown, the dust container in the dedusting system and related equipment shall be cleared.

9.2 Maintenance

9.2.1 The dust collector and dedusting device shall be inspected and cleaned regularly to prevent clogging.

9.2.2 The machine shall be shut down regularly to clean the dust outlet, which shall be enclosed in time after the dust removal.

9.2.3 The electrical equipment of the dust collector shall be inspected regularly to ensure normal operation.

9.3 Safe operation

9.3.1 When cleaning the dust outlet, the machine shall be stopped.

9.3.2 Operators shall wear safety helmets, dust masks and safety belts during maintenance.

9.3.3 When working at height, the operating platform shall be secured.

10 Cooling

10.1 Operation management

10.1.1 Before operation, the cooling equipment shall be inspected in accordance with 7.1.1.

10.1.2 During operation, the operator shall regularly check whether the product temperature is within the specified range, and adjust the parameters of cooling process as required.

10.1.3 After the molding equipment is stopped, the cooling machine shall not be shut down until the product is cooled to the specified temperature.

10.2 Maintenance

10.2.1 The cooling equipment shall be inspected regularly and the inlet and outlet shall be cleaned on a regular basis.

10.2.2 Other maintenance shall be performed in accordance with 5.2.

10.3 Safe operation

It is strictly forbidden to put hands or other objects into the inlet or outlet during operation.

11 Product Measurement

11.1 Operation management

11.1.1 Before measuring the products, the measurement equipment shall be calibrated and zeroed and the packaging materials shall be prepared.

11.1.2 When measuring the product, the range of the measurement equipment shall not be exceeded.

11.1.3 The product measurement equipment shall be kept clean.

11.2 Maintenance

The maintenance of measurement equipment shall be implemented in accordance with 5.2.

11.3 Safe operation

After the measurement is completed, all loads shall be released and the power supply shall be cut off.

12 Packaging, storage and transportation

12.1 Packaging

12.1.1 The product name, types and specifications, factory name, factory address, net weight, and executive standards shall be indicated on the product package.

12.1.2 Each batch of products shall be accompanied by an inspection certificate, which includes the product name, calorific value, moisture content, place of origin, manufacturer, date of production, seal of conformity, product standards, etc.

12.2 Storage

12.2.1 The storage warehouse must be dry, and the warehouse shall have good ventilation, moisture-proofing and firefighting facilities.

12.2.2 Appropriate ventilation spacing, as well as transportation and fire lanes, shall be reserved between finished product stacks in the storage warehouse.

12.2.3 The "first-in-first-out" principle shall be followed in the product warehouse-out. Regular inspections of product packaging intactness and appearance quality are necessary to prevent mildew and pests to ensure safety.

12.2.4 When the products are stacked outdoors, they shall be protected from rain and sunlight with the cover over and around the stacking position, and the stack bottom shall be at least 10 cm above the ground.

12.3 Transportation

12.3.1 Products shall not be transported together with inflammables, explosives, perishables, toxics or wet items.

12.3.2 The vehicles shall be covered to protect against the sun, rain, dust and moisture.

12.3.3 Products shall be handled with care.

13 Testing

13.1 Raw materials

The raw materials delivered to the plant shall be tested for moisture, impurities, etc., and the main testing values shall be recorded in detail.

13.2 Products

13.2.1 The density, moisture, ash, volatile matter, fixed carbon, calorific value, etc. of the products shall be tested, and the main testing values shall be recorded in detail.

13.2.2 Each batch of products shall be sampled for testing.
